INTRO TO PHYSICS Need to Know

SilverTip

Magnetism

by Karen Latchana Kenney

Consultant: Kathy Renfrew
Science Educator and Science Learner

BEARPORT PUBLISHING

Minneapolis, Minnesota

Credits

Cover and title page, © ShutterStockStudio/Shutterstock; 5T, © Patty Chan/Shutterstock; 5B, © Angelika Smile/Shutterstock; 7, © Ronnie Chua/Shutterstock; 9T, © taviphoto/Shutterstock and © Afiq Sam/Shutterstock; 9B, © Aphelleon/Shutterstock; 11, © valdis torms/Shutterstock; 13, © Ryan Somma/Wikimedia Commons; 14, © Prostock-studio/Shutterstock and © Kaspars Grinvalds/Shutterstock; 15T, © RHJPhtotos/Shutterstock; 15C, © RHJPhtotos/Shutterstock; 15B, © RHJPhtotos/Shutterstock; 16, © Colport/Alamy; 17, © attaphong/Shutterstock; 18, © Rashevskyi Viacheslav/Shutterstock; 18–19, © Peter Hermes Furian/Shutterstock; 21TL, © Ricwhite/Shutterstock; 21TR, © PhotoByToR/Shutterstock; 21BL, © vanitjan/Shutterstock; 21BR, © JIPEN/Shutterstock; 23, © Freedom Master/Shutterstock; 25, © Gorodenkoff/Shutterstock; 27, © ipopba/iStock; and 28, © ivector/Shutterstock.

Bearport Publishing Company Product Development Team

President: Jen Jenson; Director of Product Development: Spencer Brinker; Senior Editor: Allison Juda; Editor: Charly Haley; Associate Editor: Naomi Reich; Senior Designer: Colin O'Dea; Associate Designer: Elena Klinkner; Product Development Assistant: Anita Stasson

Library of Congress Cataloging-in-Publication Data is available at www.loc.gov or upon request from the publisher.

ISBN: 979-8-88509-223-4 (hardcover)
ISBN: 979-8-88509-230-2 (paperback)
ISBN: 979-8-88509-237-1 (ebook)

For more information, write to Bearport Publishing, 5357 Penn Avenue South, Minneapolis, MN 55419.
Printed in the United States of America.

Contents

Easy Cooking

Magnets keep a grocery list on your fridge. They also help its doors close tight. And a big magnet on the wall can even hold a knife out of the way.

Magnets keep things in place in your kitchen. But that's just the beginning of what they can do!

Listen to music blasting from speakers. Or play a game on your tablet. These things use magnets to work! Magnets are everywhere.

milk
bread
cheese

Forces at Work

Magnets make things move or keep them in place with a **force**. At their most basic, forces are pushes and pulls that cause movement.

When a hockey player slaps their stick against a puck, the force pushes the puck toward the goal. But there are also forces that move things without making contact.

We measure force in newtons. This unit is named after Sir Isaac Newton. Have you heard the story about a scientist who discovered gravity by watching an apple fall? That was Newton!

Magnetism is a force that works from a distance. It makes magnets move in two ways. Magnets can **attract**, or pull toward, other magnets. They can also **repel**. That's when magnets push away from each other. This force works on metals that become magnets.

Magnetism isn't the only force that works at a distance. Gravity can also affect objects without touching them. It can even work on objects that are farther away than most magnets.

Magnets attract or repel at short distances.

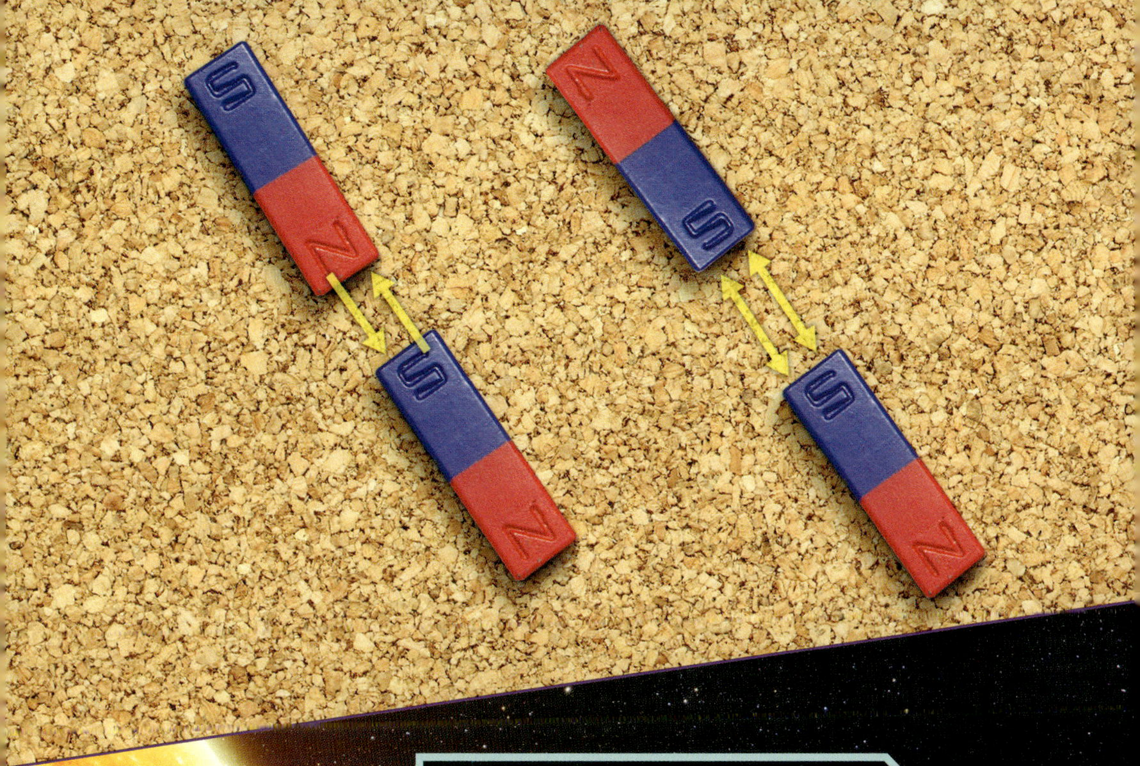

Gravity works from far away and up close.

Atomic Energy

Just how does a metal become a magnet? The secret is small. Magnetism is caused by the behavior of the teeny-tiny **atoms** that make up everything on Earth.

Electrons spin around the center of atoms. They move very fast, which creates a flow of energy.

Atoms are too small to see. In fact, it would take 500,000 carbon atoms stacked together to equal the width of a single human hair!

A Model of an Atom

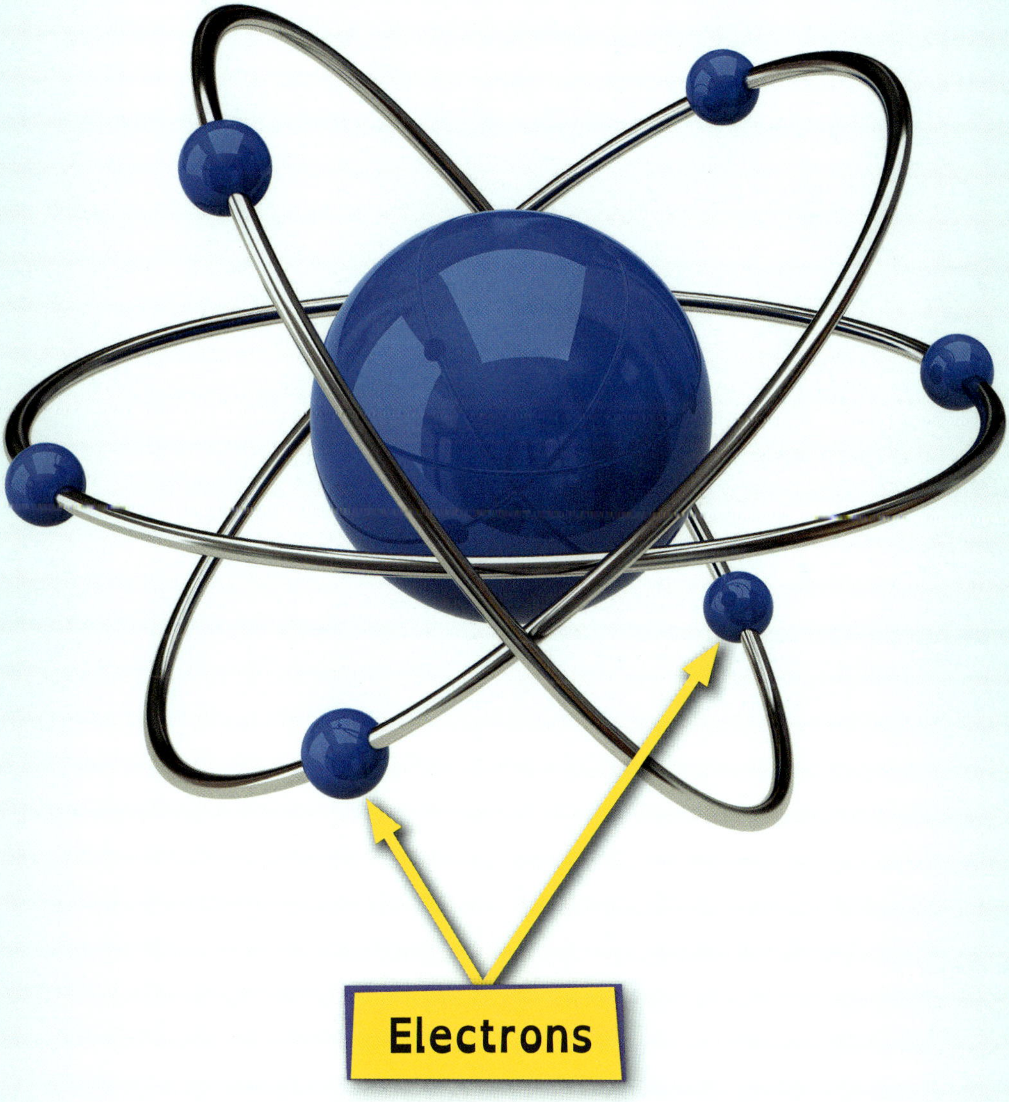

As they spin, all of the electrons in an atom create a tiny magnetic force. In most objects, electrons spin in different directions. This cancels out their magnetism.

In a magnet, the electrons move in the same direction. This makes a **magnetic field**. If another magnetic field comes close, the objects attract or repel each other.

Long ago, people first found magnets in nature. Rocks called lodestones (LOHD-stohns) attracted iron. No one knew how these rocks worked. Some thought they might be alive!

Metals into Magnets

Only certain metals can become magnets. Their electrons move in a way that is more likely to make a magnetic field. Magnetic metals are iron, cobalt, and nickel. Some rare earth metals are magnetic, too. A mix of these metals can also be put together to make magnets.

You may have some magnets in your pocket right now! Most smartphones are made with parts that have rare earth metals.

Iron

Cobalt

Nickel

Powerful Poles

All magnets have two **poles**. Magnetic force flows out of the north pole. It flows back into the south pole. This is what creates the magnetic field around a magnet.

Magnetic forces push two similar poles apart. They pull opposite poles together.

One of the first scientists to study magnets was William Gilbert. In the 1500s, he did experiments with lodestones. He came up with the term *magnetic pole*.

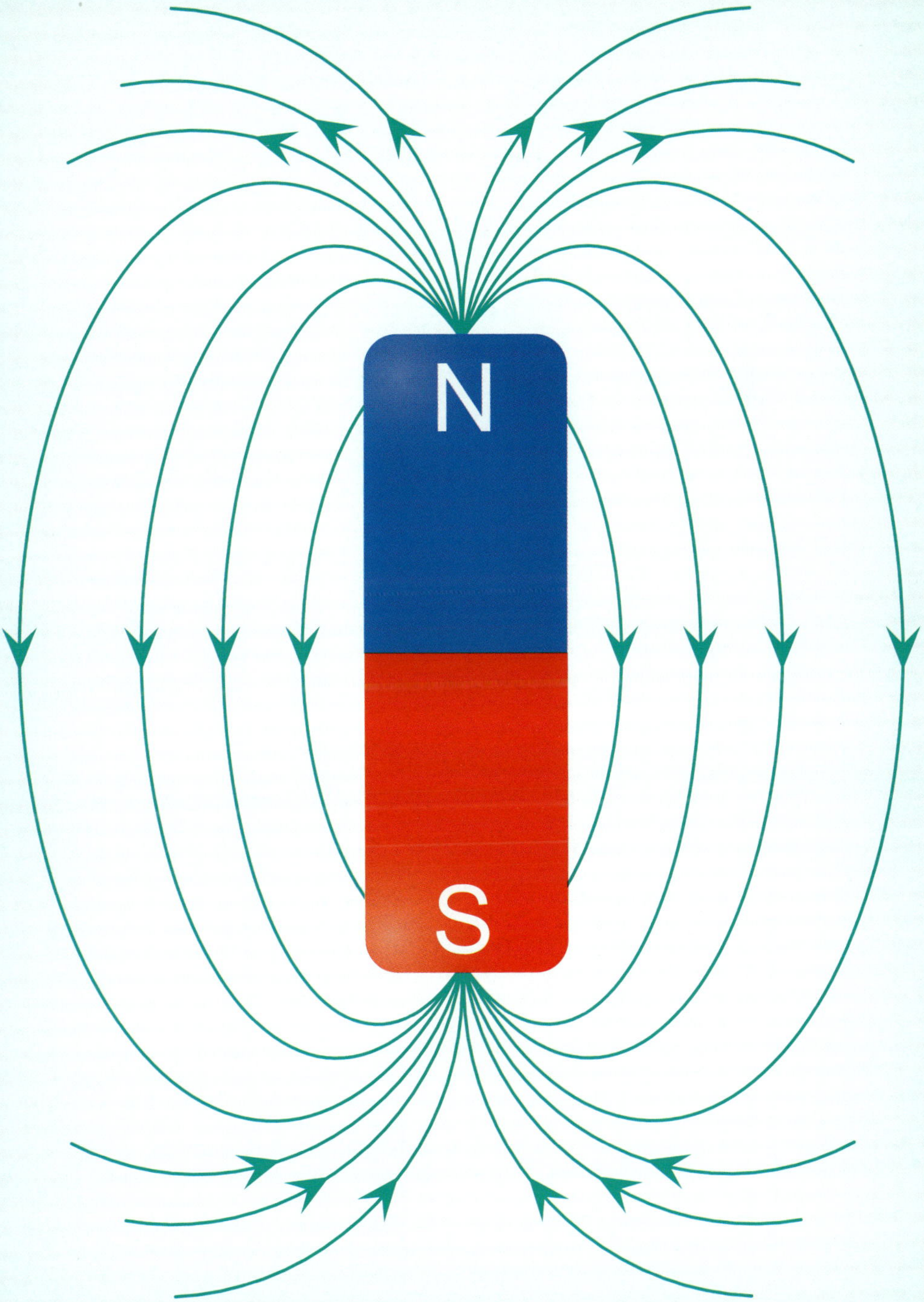
N
S

Just like a magnet, Earth has two poles. That's because Earth's **core** is made of hot, liquid iron. This metal makes a magnetic field around the planet. Its force helps stop harmful space objects from slamming into Earth's surface.

A compass relies on Earth's magnetic field to tell direction. It has a magnet that is attracted to Earth's north pole. This magnet turns to always face that direction.

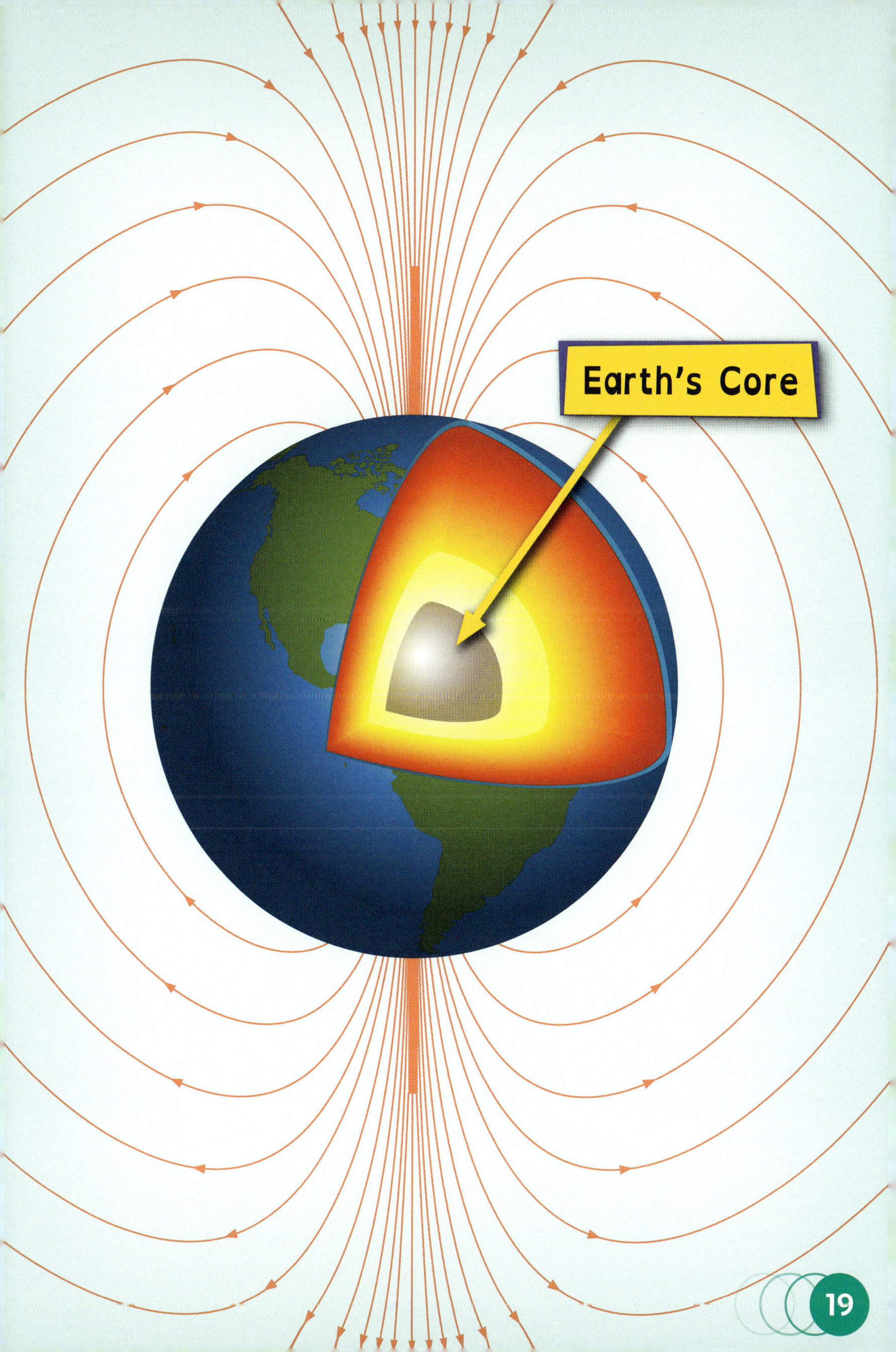
Earth's Core

Always On

Some objects are always magnets. They always have a magnetic field. These are called **permanent magnets**.

The force from these magnets is used to power many machines. Electric motors in microwaves use magnetism to work. These magnets are also inside computers, televisions, and speakers.

> Some permanent magnets can be found in nature. They were made when lightning hit a mineral called magnetite.

Sometimes On

Some metal objects can become magnets for a short time. They are **temporary magnets**. These metals are magnetized when they are near a permanent magnet's magnetic field. The metals slowly lose their magnetism after the permanent magnet moves away.

Many kinds of metal objects can become temporary magnets. Rub a permanent magnet against a metal paperclip, pair of scissors, or iron nail. The metal becomes magnetized.

Temporary magnets can attract other metals.

Another way to make a temporary magnet is by using electricity. Sending electricity through iron makes a very strong magnet. The magnetic field stops when the electricity is turned off.

Electric magnets are in trains and construction machines. Hospitals use them to create images of the body.

An MRI machine uses an electric magnet to let doctors see inside bodies. It changes how parts of atoms line up inside the body. Then, the atoms send out signals that we use to make pictures.

A person lies inside the tube-shaped magnet of an MRI.

Magnetic Research

We already use magnetism in many ways. And we're looking for even more uses.

Someday, we may even be able to use high-powered magnets to make electricity for our homes and cities. What does the future hold for this force? We'll have to wait and see.

Where is the strongest magnetic field in the world? It's at the National High Magnetic Field Laboratory in Tallahassee, Florida. This field is stronger than 455,000 fridge magnets.

Magnetic Fields

Magnetic force always travels out of the north pole. It travels back in through the south pole. This creates a magnetic field.

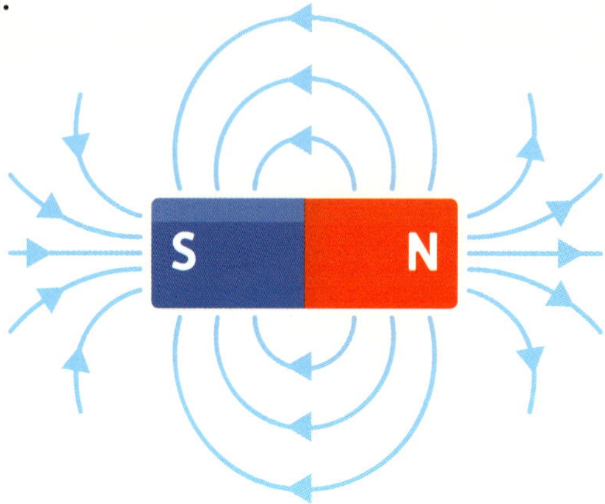

Magnets attract at opposite poles.

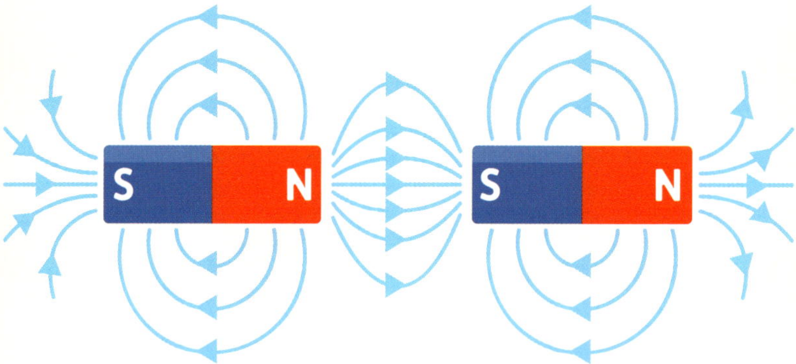

They repel each other from their similar poles.

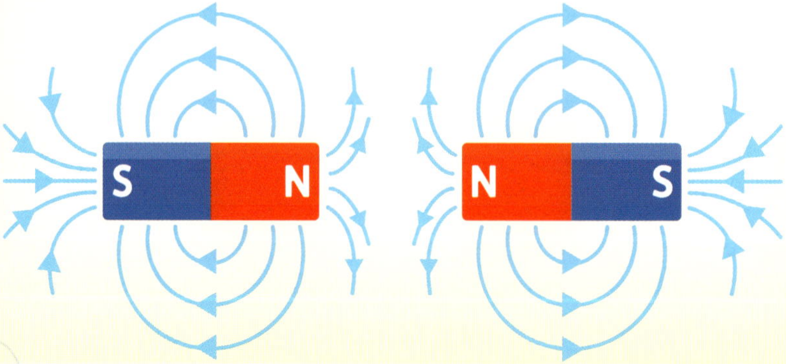

SilverTips for SUCCESS

★ SilverTips for REVIEW

Review what you've learned. Use the text to help you.

Define key terms

atoms
magnetic field
permanent magnet
pole
temporary magnet

Check for understanding

How does metal become magnetized?

What happens at a magnet's two poles?

Describe the difference between temporary and permanent magnets.

Think deeper

How do magnets make your life easier? Think of a few examples.

★ SilverTips on TEST-TAKING

- **Make a study plan.** Ask your teacher what the test is going to cover. Then, set aside time to study a little bit every day.
- **Read all the questions carefully.** Be sure you know what is being asked.
- **Skip any questions** you don't know how to answer right away. Mark them and come back later if you have time.

Glossary

atoms the tiny building blocks that make up every substance in the universe

attract to pull together

core the center of Earth

electrons tiny parts of an atom that contain an electric charge

force a push or a pull that causes movement

magnetic field a force around a magnet that pushes or pulls other magnets or metal

permanent magnets magnets that always have a magnetic field

poles the two points of a magnet where the magnetic force enters and exits

repel to push apart

temporary magnets magnets that have a magnetic field only sometimes

Read More

Gianopoulos, Andrea. *The Attractive Story of Magnetism with Max Axiom, Super Scientist: 4D An Augmented Reading Science Experience (Graphic Science 4D).* North Mankato, MN: Capstone Press, 2019.

Miller, Derek. *The STEM of High-Speed Trains (The World of STEM).* New York: Cavendish Square, 2021.

Owen, Ruth. *Forces & Magnets: Let's Investigate (Science Essentials).* Minneapolis: Ruby Tuesday Books, 2021.

Learn More Online

1. Go to **www.factsurfer.com** or scan the QR code below.
2. Enter "**Physics Magnetism**" into the search box.
3. Click on the cover of this book to see a list of websites.

Index

About the Author

Karen Latchana Kenney is an author from Minnesota. Magnets hold up photos of her favorite people and places.